WHAT IS TIME?

David Carson

What is Time?
@David Carson

A Carson Group Book
ISBN: 978-1-300-65080-5

carsonskc@gmail.com

Science

If everything were static and nothing ever changed, perhaps time would not exist. After all, isn't time used to measure change, and if there is never any change, why think of time in the first place? Time measures change.

Would time exist if nothing ever changed? Well, perhaps it would and perhaps it wouldn't. This is a philosophical question which is far beyond the scope of a brief essay like this claiming no special authority or insight. However, it appears time is central to both physics and engineering although many doctors of physics claim time doesn't exist, that time is an illusion. Some of the ancient poets such as Omar Khayyam are more definite about the nature of time. According to Mr. Khayyam, time flies like an arrow, though of course that doesn't mean that time is an arrow. The great authority of our time, Dr. Anthony Fauci, is silent on this subject which is not of great importance like his normal subjects concerning life and death, vaccines, the proper wearing of mussels or masks, and the necessity of having sticks jammed up the nose in an experimental way. However it is very satisfying to have an authority like Dr. Fauci who represents science. He says so himself.

But what is science? Is it the piling of one hypothesis on top of another, each hypothesis subject to skeptical investigation in the same way the ancients thought of the universe as four great world-bearing elephants standing on top of a huge turtle piled on top of another turtle, with turtles all the way down? The turtles all the way down precedent or argument was set by Judge Alioto of the Supreme Court, another great authority like Dr. Fauci.

The problem with the idea of science as a bunch of falsifiable hypotheses subject to skeptical investigation is one of authority. How can science be taken as an authority in anything important such as politics or law if it is always subject to skeptical investigation. Certainty doesn't allow questions. Authority doesn't allow questions. Dr. Fauci doesn't allow questions. If Dr. Fauci represents science he represents science as certain facts that can't be questioned. A science that is certain, that is frozen in place in its incredible majesty and power can be a foundation for law itself, for society itself. A science given over to corrosive skepticism can be the foundation for nothing other than anarchy and chaos, for dissolution and even the denial of the most divine authority. But science as derived from God himself, as a description of the mind of God, is both certain and a solid foundation for law and order, and, by the way, for continual lockdowns. Lockdowns were so good at ruining independent businesses and promoting cartels and monopolies that they may become part of the new science, the new Fauci-Science. The mind boggles at the possibilities. Next could be a climate change lockdown featuring fuel rationing, closing of borders, and widespread fuel police to complement the good works of the pandemic police. Finally we could have universal lockdowns to catch dissidents, the kind of traitors and terrorists that think Dr. Fauci is a scoundrel if not an idiot too. Terrorists that hate tyranny could not be more dangerous to the establishment, to the Faucis and those who stand as beacons on top of our great civilization. Those in authority must be respected even though they may have their faults.   For example, Dr. Fauci must be respected even though it appears he was instrumental in funneling money into gain-of-function research into labs all over the world, even labs in China. It might appear that such research breaks the treaties banning biological weapons, which is only a superficial and stupid thought. Why should any principles or morals or treaties of any kind restrict the leaders of our great and exceptional country. Did J. Edgar Hoover, another hero in the mould of Dr. Fauci, ever let principles or morals restrict his COINTELPRO programs, his other skullduggeries, or his dancing around in his pink tu-tu. Of course not!!

It is not a coincidence that medicine and public health medicine in particular are at the center of many of our current problems. Medical journals in the U.S. have become notorious for printing advertisements for some drug as scientific research. Sometimes the scientific results can not be duplicated by other, independent labs. Either reality changes from one lab to another or we have made a great breakthrough in science with the understanding that the truth is relative and variable. What is sometimes true is sometimes false. Sometimes one plus one is two; but sometimes one plus one is really three or whatever. The idea that one plus one is always two is a fallacy of white racism or something of that sort; it's hard to say since truth is so relative and variable. No doubt, in certain circumstances, voodoo is the absolute truth. Truth is all about the Benjamins, at least in those parts of the world where the ineluctable charm of the Benjamins is recognized.

The nonsense published in medical journals as science was a precursor to our current phony pandemic, and the attempts to build biosecurity totalitarian states around a respiratory disease. This has to be a great innovation in statesmanship and geopolitics worthy of a Bismarck or a Kissinger or George W. Bush or even of the great Barack Obama himself. In many countries the local worthies have enjoyed themselves making up all kinds of arbitrary rules around the pandemic. They dictate spaces of six or seven or three or whatever meters or feet between victims, or do I mean citizens? They dictate closing of all non-essential businesses, and then define essential in an entirely arbitrary way, according to their own whims and the orders of their rich donors and cronies.

Juicy alarms abound, virtually anything can be used to strike fear and suspicion into the minds of the dutiful citizen. Perhaps the juiciest is climate change, and the horrible things all of us have done to Greta Thunberg and all the rest of the children. Climate change may also demand lockdowns and complete totalitarian control of a process of depopulation and the further concentration of capital in a few, responsible hands. A restriction and rationing of fuel could easily collapse cities built around cars and trucks.

No doubt the amount of trace amounts of carbon dioxide and methane have increased in the atmosphere due to industrial expansion; and no doubt this appears to give a warming effect. However, there is no clear idea of where exactly the planet is with respect to its cyclical ice ages; and there is certainly no clear idea if it would be possible to reverse climate warming by locking down that part of the planet under oligarchic control. Are we really that powerful, or are we just that arrogant?

Reality

"It's amazing how far a little common sense goes and it's also amazing what happens when there's no common sense at all."

Lynn Turner, SEC accountant

If there is no common sense, and there certainly appears to be very little if any at all, then there is no idea of the nature of reality. The man with no common sense has little idea of reality either; he is a perfect slave willing to accept his master's idea of reality, if the master has any idea of reality himself. But what is a master?

A master can be anything or anyone that is accepted as an authority, an expert. And a man with no common sense will accept virtually any fool as an authority. A country with no common sense is a disaster; overrun by confidence men and women such a country stumbles from one calamity to the next, from one humanitarian bombing to using its population as lab rats in medical experiments. History is a dreary chronicle of such countries, empires, or principalities and their murders, injustices, and outrages; a history of the blossoming of stinking flowers of tyranny and their eventual decay.

But we have discovered an antidote: science and its methods. Science is accepted by most as our means of seeing reality as it is, not as seen by some confidence man. Unfortunately, science is not a panacea for all our problems with reality. For example, the idea that Dr. Fauci could get away with representing "science" is an indication of how little common sense exists in our media and elite circles. In other words, Fauci corrupts the idea of science, and turns it into nothing more than another authority that must be accepted with no questions asked. This is an outrage. One of the bases of science is the ongoing questioning of all its hypotheses and assumptions, all of which can be falsified by clear-sighted observation of reality, an objective reality that exists outside of any man or even any particular species.

Of course Dr. Fauci is not the only person to corrupt "science"; or rather, the word "science". Political "science" and social "science" are both forms of questionable philosophies that muddy linguistic waters to aggrandize themselves. One of main differences between the social sciences and true science is the social sciences are not cumulative in the same manner as mathematics, chemistry, or biology. That has to be the case given the nature of our political systems that use the social "sciences" to justify their various actions and crimes. A change in regime often leads to sea changes in political philosophies.

This just scratches the surface of our current situation, our descent into an age of corruption, collapse, and chaos. Even science and the scientific method are attacked as malignant expressions of white racism and white supremacy; a deep explosion of hatred and anger. According to this point of view racism is the only possible reality, hatred is the only possible reaction to this reality. Hate must have a target; but almost any target will do, even the scientific method. Another response to our vortex of corruption, collapse, and chaos is the search for certainty and meaning; for anything as a life raft for the drowning mind. Authority is the only answer, and virtually any authority will do, even someone like Anthony Fauci with his stream of contradictory commands, his kaleidoscopic realities that seem to change at random.

The philosophies at the intersection of the hard and soft sciences are particularly interesting now. Economics and medical science both appear to be in this strange category. This is not to say that economics and medical science are not very useful and interesting parts of philosophy; but they are not the same as mathematics or chemistry.

Economics is another philosophy that fails to be cumulative, that changes with the political and social winds. Currently, at least in the U.S., modern monetary theory is very much in fashion. It justifies money printing by the government on a grand scale, at least for those countries which have their own currencies. The idea that money printing easily debases and destroys a currency as was the case in Weimar Germany, Venezuela, Argentina, and so on is totally discounted as nonsense from those who simply are not smart enough to understand the subtleties of the theory.

As pointed out by Michael Hudson, modern economics ignores the damages done by a predatory creditor class that insists on repayment of debts regardless of circumstances. There is nothing wrong in the eyes of modern economics with any kind of rent seeking, and public-private partnerships are seen as benign rather than as a looting of public assets.

Modern economics appears to be declining, not advancing in any sense. Yet sophisticated statistical and mathematical methods are used in economic papers to lend an appearance of deep and advanced thought.

In addition the ongoing collapse of the international dollar system is ignored. The connections between economics, geopolitics, sociology, etc. are also ignored. Philosophy splinters into all kinds of pieces and is thereby destroyed.

Currently medical reality is all about getting your shot. It is a huckster's reality centered around the lucrative vaccine program. Any doctor who proposes any cures other than the expensive shots is at least ignored if not canceled entirely. Dissenting voices are heavily censored by the media platforms that do the government's dirty work and allow the politicians their sacred plausible deniability. Anyone who points out the cost of the shots is shouted down. The shots are free!! Of course the shots are not free; they run around so much a jab to the government; but since the federal government is already grossly insolvent it really makes no difference what the shots cost. Reality is altered to make the shots necessary to life itself. Only a thorough going media monopoly is capable of creating such a bizarre reality in which an experimental jab becomes a necessity.

Once everyone is immersed in an insane reality, how is it possible to make any progress understanding the other reality; the reality that exists on its own without the help of politicians, hucksters, academic con-men on the make and on the take, or the access one particular species has to a very narrow part of the electromagnetic spectrum?

How is it possible to discuss the nature of time in a thorough way?

Old Time Problems

Time grew out of the shadows of a star that obediently circled around us everyday. What a surprise it must have been when it was discovered we circled the sun rather than the other way around. As a matter of fact, Copernicus' heliocentric system was considered heretical by some of the most important officials who had devoted their lives to shielding everyone from their ignorance with their own invincible and specialized ignorance. The Ptolemaic system of epicycles used to explain the motion of what turned out to be other planets had to be junked. What an outrage for dedicated astronomers! However, shadows cast by the sun could still be used to measure the flow of time through sunny days. Also, the old calendars could still be used. It was not a complete disaster.

As time passed, machines advanced even as humanity stagnated, and it became possible to read the exact zenith of the sun over a particular spot on Earth. In the nineteenth century the sun's zenith was taken to be high noon in each locality. Each local time belonged to each

town; and many times were therefore different. Of
course it has always been known that the sun has been
up for hours in Maine at dawn on the left hand coast. In
this golden age, each town had its own time, each bank
it own currency, and every railroad its own time dictated
by its headquarters's time.

It was an actual advance in understanding, a very rare
thing in our history, that we traveled around the sun
rather than vice versa. Of course it came late in our
career as a species, late in the career of homo sapiens;
or should we call ourselves homo moroni or homo
imbecili in honor of our violent and fairly idiotic history?
Of course wisdom couldn't exist without stupidity;
therefore we can perhaps forgive ourselves.

The new understanding of our whirling orbit around the
sun led to both the ideas of local solar noon and to a
precise calendar. Unfortunately before the age of
cartels and monopoly each railroad's different time led
to confusion. A Canadian, Sanford Fleming, designed a
system to divide the planet into 24 time zones to
correspond to the 24 hours in a day, and introduced the
idea of an international date line to prevent us from
repeating the same day ad infinitum. This system was
adopted after much deep thought about where the

beginning or prime meridian would be. Since most ships at sea used Greenwich time; the railroads and localities accepted the same prime meridian. The Greenwich prime meridian is one of many relics of the British Empire scattered all over the planet such as the British Commonwealth, the C.I.A., American Ivy League colleges, Indian railways, etc.

The idea of each locality having its own time based on the apparent position of the sun was extended or embellished over a hundred years ago in the theory of special relativity to include every observer. In other words, each person, depending on his position and movement, had his own version of time according to this theory.

In order to look at the various ideas of the nature of time, some physics must be considered. This is unfortunate because modern physics seems to be afflicted with the same disease as economics and some of the other social sciences; questionable if not wrong applications of mathematics. In the case of some physics the misapplications are so bad that they are "not even wrong," a famous phrase Wolfgang Pauli would use to humiliate floundering students. "Not Even Wrong" is the title of a book by a mathematical physicist, Peter Woit, discussing the currently fashionable string theory used in particle theory and to fill entire volumes of the

Physical Review, one of the principal publications of mainstream physics.

Eugene Wigner, an accomplished mathematical physicist and one of the founders of quantum theory, wondered why mathematics should be so useful and essential to a description and understanding of physical reality. Wigner was one of the men responsible for applying group theory to particle theory, a process called Gruppenpest by the sharp tongued Pauli. There is no clear reason why group theory should apply to tiny physical particles; or at least that was probably the opinion of Pauli who passed from the scene many years ago to some theoreticians relief.

Unfortunately, there is also no clear reason why eigenvalue spectra, a part of the theory of differential equations, should give the physical values of measurements on these same tiny particle. This is called the measurement problem, a problem that is not ever mentioned in polite company in the same manner that the emperor's lack of clothes was never mentioned in polite society. Most of those who take up this problem are not initiates in the high physics priesthood, and are therefore ignored.

There is no clear reason why large objects should follow

classical mechanics while very small systems follow quantum mechanics. This is called decoherence theory, which invokes whole systems of particles to achieve the coherence of a large object or system.

Incidentally, quantum theory or mechanics is the primary theory of modern physics and commands the attention of many more scholars than relativity or cosmology or any other part of modern physics.

Considering the questionable applications of group theory, eigenvalue mathematics, and many-particle theories in decoherence, it should be no surprise that conventional physicists have some very surprising ideas about time itself. According to Julian Barbour time does not exist; the passing of time is an illusion. Perhaps he believes old age and death is an illusion also. Kurt Godel, who was primarily a logician rather than a physicist, believed he had proved the nonexistence of time by constructing a universe in which Einstein's equations of general relativity yielded the possibility of time travel. Godel's universe was very strange, if not physically impossible. Of course physical impossibility is not something that would stop a devout logician. And then there's the arrow of time. Physicist have a lot of trouble with the arrow of time since it's not obvious, at least to them, why time should flow forwards rather than backwards. Philip K. Dick, the master story-teller whose paranoid fantasies seem to be a template for the future, wrote an entire novel in which time turned around and began to flow backwards. The story is told from the point of view of a grave-digger whose job is to dig up the people who are coming back to life.

More Problems

The physicists have difficulty with time even though many of their descriptions of physical systems claim to trace changes in time. Time is not really understood; but then, there are a lot of things that are not understood. According to Alex Unzicker, a German who has acquired the mantle of the boy pointing out the nakedness of the emperor, the Standard Model leaves many problems unanswered. The Standard Model was proposed over half a century ago and claims to explain the universe (except for gravity). As Unzicker points out the Standard Model does not compute masses of fundamental particles, their lifetimes if they are not stable, the infinities in electrodynamics, the origins of gravity, antimatter, and spin, and the nature of space and time.

Unzicker goes even further. He claims many physicist are using mathematics to obfuscate their ignorance and to gild swamps of fantasies in the same manner some economists use mathematics to hide their prejudices and ignorances in favor of their wealthy donors. In this sense physicists appear worse than any of the phony Nobel Prize economists since no wealthy donor really gives a damn about the lifetime of a radioactive particle or the mass of an electron. The donors (militaries, governments, etc.) only seem to care about the physicists' bombs; the ability of blow everything straight to hell.

Unzicker could also be seen as Don Quixote, tilting with windmills to no avail. The emperor and the windmill are conventional physics, which used to hold the position in academies as the emperor. Perhaps it has lost its position in the U. S. to American Studies, Gender Studies, or Black Studies; disciplines that will put up with no questioning or criticism and police that which is allowed and that which must be canceled and censored in the name of identity politics and anti-racism. However, none of this doesn't mean Unzicker is not correct.

In particular, Unzicker believes it is a mistake to conflate time with the spatial dimensions as is done in Minkowski's space, which is designed to incorporate the constancy of the speed of light in a four dimensional space. Minkowski space is really quite strange since the time dimension requires multiplication by the square root of minus one and a constant that gives the speed of light. Unzicker believes time should be treated separately from length, height, and width because it only flows in one direction. The spatial dimensions don't flow at all and allow backward and forward motions at will.

Unzicker, and others have formulated general relativity in three dimensional spaces with a variable speed of light to agree with Einstein's statement that the motion of light depends on its position in the gravity field and that special relativity is only correct in the absence of gravity.

The mathematics of differential geometry, Riemann geometry, and tensor calculus were developed before relativity theory. Mathematicians generalize their ideas and results as much as possible; and it's easy to extend differential geometry and tensor calculus to any whole number of dimensions. Therefore it may have been natural for Minkowski to try to incorporate the results of the Michelson-Morley experiment in a 4-space; even though the inclusion of time requires the odd (from the point of view of physical phenomenology) factors of the square root of minus one and the speed of light. Initially Einstein did not put special relativity in Minkowski's space; however he used four-space in his later development of general relativity in a four dimensional metric space probably due to the influence of mathematicians and the prior existence of differential geometry. The space metrics of the resulting space indicate the curvature of space due to the presence of matter; but the physical meaning of the elements of the metric involving time, the fourth dimension, are more obscure.

Newton seems to have developed his theory of gravity and the mathematics of calculus in tandem, as needed to describe the ellipsoidal orbits of the planets observed by Kepler. Calculus, at least in somewhat its current form, did not exist before Newton and Leibnitz. Currently the fashion in theoretical physics is to apply finished mathematics to physical problems and hope all the necessary assumptions actually describe the physical situation at least in some reasonable approximation.

There are other unsolved problems not mentioned in Unzicker's criticism of the Standard Model, which may involve our understanding, or rather our lack of understanding, of time. Our local galaxy and other galaxies appear to be rotating far faster than they should be rotating given the amount of mass visible in them. The usual solution is the postulation of dark matter that can not be seen, but which is still able to effect galactic gravitational fields and cause the rotation anomaly. This additional matter can not be seen, so it's called dark matter. This is reminiscent of the addition of extra epicycles in Ptolemaic astronomy to account for the unusual and anomalous movements of other planets as seen from the Earth. The idea that everything might not be rotating around the Earth was never entertained or was even considered heretical. The use of a galactic clock based on the rotation of the Milky Way throws our present system of measuring time in question. Therefore dark matter is considered a better solution.

Also, it is surprising that time is treated in the same way in both quantum and relativity theory; the two theories treating the large and the small. It is also surprising that there have to be two different theories treating big things and small. It is even more surprising that the two theories don't appear to be compatible in any way; in other words, it is hard to see now one theory merges into the other in mesoscopic objects, objects halfway in size between an electron and a galaxy. What is a classical object, one that obeys relativity and Newton's physics in an approximation? Both superconductors and superfluids can be fairly large yet seem to have some of the properties of quantum objects. In quantum field theory, an extension of quantum mechanics purportedly covering very small systems with many particles that may appear and disappear in various atomic processes, the number operator is a global entity describing the system and its entire surroundings. The number of particles in the system may vary depending on the motion of the observer's co-ordinate system in a curved space. And in quantum theory in ordinary space, the wave function is taken to describe the entire system to be described out to infinity. The wave function is global; yet in Einstein's equation of general relativity all the quantities refer to fields at each point in space and time. In other words, the metrics, curvatures of space, and stress tensor are microscopic point-to-point mathematical objects. Yet relativity is taken to describe large objects. Isn't this strange?

Time Everywhere

A key advance of relativity is the idea that our means of seeing, or communication, needs to be incorporated in our idea of time. Imagine a crowd watching a huge clock with only one hand that rotates very rapidly. Each person would see the hand of the clock in a slightly different place depending on his location in the crowd since it would take light reflected off the hand different times to reach the observer. Newton solved this problem by using what might be called universal time. Now is now everywhere in the universe; there is no need to use any kind of synchronizing signal such as light or sound.

The difference between Newton and Einstein's point of view of time might seem to be nothing more than a minor quibble until it's applied to an atom radiating light. Einstein used conservation of energy and the Lorentz transformations of space and time to come up with his famous equation energy equal mass times the speed of light squared, describing the energy in the rest mass of the radiating atom. His equation is taken as a universal law, applying everywhere in the universe, even though the properties of light have only been directly observed here on Earth. This kind of thing is a very large generalization. Whether or not it is justified is another question altogether.

Are we dealing with universal laws or just rules of thumb that happen to apply in some cases? It is painfully obvious that many of the universal laws proposed by eminent social scientists are no more than rules of thumb, at best. It is painfully obvious in their case because their disciplines are not cumulative, but seem to depend on whatever is fashionable and whatever pleases powerful donors of stipends. How about the hard, cumulative sciences, how about physics, chemistry, and mathematics? According to some authorities in the U.S. the hard sciences are simply constructs of racism and white supremacy, fairy tales meant to confuse and impress gentlemen and ladies of color. Although skepticism is honored by the best scientists in any discipline, hard or soft, idiocy is rarely honored anywhere. This really is a universal law.

According to Lev Landau, an eminent Russian physicist, cosmologists are often wrong but never in doubt. To what extent can his statement be applied to others?

It seems to be virtually impossible to be skeptical of mathematics and logic. We must honor these disciplines; without them we would be lost in a morass of superstitions and illusions produced by powerful political magicians to manipulate and control their dummies. We would be back in the Dark Ages.

Quantum theory, the current center of physics, uses time as a parameter, separate from the other dimensions. Usually, discussions of the measurement problem, which has dogged quantum theory since its beginning, do not involve time. In quantum theory, at low velocities, it is silently assumed Newton's time will do to measure change. But what about a relativistic quantum field theory in which ghost particles can be seen by an observer in an accelerated frame? What if the observer doesn't use light as a synchronizing signal as we do?

We use light as our synchronizing signal because it is natural to us; it is our intuitive choice. We hunt with light. However, porpoises, bats, and whales hunt with sound, with sonar. Wouldn't it be natural for a whale Einstein to use sonar to construct his version of relativity? Light is of not much use in the deep dark expanse of the great oceans that cover most of the

Earth.

What if an extraterrestrial Einstein of whatever species also used light as his primary signal too, but it was not our light? Or what if the nature of light is changing with time here and everywhere? This brings up the ancient problem of whether or not nature's constants such as the speed of light and the strength of the gravitational field are really constant; and if they are constant, why are they constant and why do they have their particular values.

The curvature of space in general relativity is given by a space metric, a tensor quantity that describes the space. A special case of these mathematical spaces is Minkowski space, sometimes called flat space since there is no curvature. All of these spaces include the speed of light in those parts that refer to time elements. The conversion factor, the speed of light, can not be avoided unless it is set equal to one. The argument for this extraordinary assumption is that one can use any arbitrary system of units as is convenient; so why not use a system where the speed of light is taken to be one. This is an obfuscation as far as the physical situation is concerned, although it may be very pleasing to those primarily interested in the mathematics of differential metric spaces. The speed of light is very large as far as our world is concerned, and furthermore, in the diagonal time element of the metric it appears as the speed of light squared, an even larger number. Any variation in this number could therefore lead to large physical effects.

Any variation in the fourth diagonal element of the metric due to a variation in either the speed of light or any other cause might offer an explanation of the measurement problem. Whenever a measurement is made of a quantum system such as an atom a definite value is found in our classical world; that is the world of things our size here in our small place in the universe. In other words, the machinery of partial differential equation theory is conjured into place and use. There is no accepted explanation for this magic. Perhaps our measurement of a quantum system forces it into a kind of time where the correct synchronizing signal is our version of light. Perhaps the variations in the positions and velocities of a very small particle are really variations in the atomic or quantum time scale. In order to avoid any accusations of false modesty, perhaps a quantum theory cast in this form is capable of union with relativity theory, and perhaps physics could finally offer a philosophically consistent theory of the physical world in all its sizes and kinds.

The current quantum theory is a philosophical failure, basically because of the measurement problem. In addition, current physics is chock full of what are kindly called "anomalies" from dark energy and matter to the failure of the CERN accelerator to find any supersymmetric particles predicted by particle theory. Meanwhile the Standard Model of particle physics, put in final form in 1967, remains untouched even though it appears to be nothing more than a complicated version of the Ptolemaic epicycle theory of antiquity. Only a German high school physics teacher, Alex Unzicker, dares point out the emperor's nakedness.

And what is time? Time is used to measure change; but what kind of time should be used for each change? Should it be Newton's time, relativistic time, or a time based on some other synchronizing signal than our version of light?. We need a deeper and wider understanding of time.

www.ingramcontent.com/pod-product-compliance
Lightning Source LLC
Chambersburg PA
CBHW081347180526
45171CB00006B/619